# PAWNS IN A GAME: A FLAT, MOTIONLESS EARTH

Written By: Brian "Big-Country" Bender

1 Pawns In A Game

A quote from one of my favorite books; a book that began my journey to awakening.

"None Dare Call It Conspiracy" by Gary Allen

*"However, the most effective weapons used against the conspiratorial theory of history are ridicule and satire. These extremely potent weapons can be cleverly used to avoid any honest attempt at refuting the facts. After all, nobody likes to be made fun of. Rather than be ridiculed most people will keep quiet; and, this subject certainly does lend itself to ridicule and satire. One technique that can be used is to expand the conspiracy to the extent it becomes absurd. For instance, our man from the Halls of Poison Ivy might say in a scoffing arrogant tone, "I suppose you believe every liberal professor gets a telegram each morning from conspiracy headquarters containing his orders for the day's brainwashing of his students?" Some conspiratorialists do indeed overdraw the picture by expanding the conspiracy (from the small clique that it is) to include every local knee-jerk liberal activist and government bureaucrat. Or, because of racial or religious bigotry,*

*they will take small fragments of legitimate evidence and expand*

*them into a conclusion that will support their particular prejudice,*

*i.e., The conspiracy is totally "Jewish," "Catholic," or*

*"Masonic". These People do not help to expose the conspiracy,*

*but, sadly play into the hands of those who want the public to*

*believe that all conspiratorialists are screwballs."*

*You will notice that after every chapter I purposely left a page blank. This is where I gave you, the reader, the option to make notes on the chapter from above. Use these areas to expand your knowledge and to further review the contents of the chapter in your own time.

5 Pawns In A Game

# CHAPTER ONE

## The Game of Chess

*"Some people think if their opponent plays a beautiful game it is ok to lose. I don't. You have to be merciless."*

-Magnus Carlson

The game of chess is a complex one that requires skill and mental stability. I enjoy the game of chess and the more I think about it, we all play the game of chess at least once in our life. I don't necessarily mean that every person will sit down and play a game of chess against an opponent, I mean that there is a great game of chess that is being played every single day. On every street corner in every city in every town of every nation. To those of us in the "know" we realize that we are on the metaphorical chess board. If you haven't realized you are playing a game of chess, let me welcome you to this game! After reading this book you and perhaps everyone around you will begin to understand that you are all also in this game of chess and this game is rigged against us.

**The Pawn:** The Pawn is moved to open up the board. They allow for other higher-ranking pieces access to attack. It is the easiest piece to move as well as the most "worthless" on the board. Forward and backward are the pawns' only movements; it may attack but only at a diagonal movement. Now obviously the pawns do not move by themselves,, someone is controlling all of the pieces with one goal in mind, victory and control. Pawns are the workers, the ones that are sacrificed with no real consequence; collateral damage if you will. Who in society might be the pawns you may ask? Well, that would be us. You, me, your co-worker, your pastor, your family. We are the collateral damage to the elites.

**Mid-Level Pieces:** Bishops, Rooks, and Knights are all pieces that I would consider mid-level. Who in society are these pieces associated with? The Media, the celebrities, the athletes, the social media influencers, and self-proclaimed spiritual gurus. Their purpose is to set up the attack. The attack, well we'll call this the narrative, after the pawns have been moved into their appropriate places. Notice how on the board you have two pieces on either side of the king and queen. One representing both sides of the aisle. Let's call the left mid-level pieces the "representatives" which we will call the liberal side; CNN, MSNBC, the Democrats, etc., and the right mid-level pieces the "representatives" of the conservative side; FOX, Breitbart, the Republicans, etc. You may stop and ask yourself, "Wait that doesn't make sense. Shouldn't these pieces be on opposing sides of the board in this game, versus the same side separated by the King and Queen?" Well no they shouldn't because as though they appear to be engaging in battle against one another on the media screens that consume every facet of our daily lives they are very much on the same team. They always have been and they always will be. If the pawn were to look behind them they would see what would appear to be the left versus the right split on opposite sides of the king and queen pieces but to the unknowing pawn, they are still on the same side of the chess board. The illusion has been set into play perfectly by the mid-level pieces creating so much chaos it's hard to follow what is real and what is fiction. Who is a good guy, who is the bad guy, and who should I trust? Will this candidate save me and my family? All the while they are taking orders on behalf of the king and queen.

**The King and Queen:** That leaves us with two of the most important pieces on the board and yet still not the strongest. You may be thinking that the king and queen are the strongest pieces on the board because they seem to be the movers and shakers. If so, you are forgetting a significant part of this picture I am painting for you. The king and queen may have the most power visible; The president, the congress, pharmaceutical companies, bank CEO, etc. They still aren't the ones making the pieces move. Someone or some group of people are the ones playing the game, making decisions in the world, manipulating certain narratives so their plans will never be exposed. Have they always been bloodline? Are they in control due to Aristocracy?

"Give them bread and circuses and they will never revolt." Keep the population at each other's throats so we will never awaken from this matrix that we are trapped in. What I'm getting to here is no matter where you place yourself on the societal ladder, you are just a piece on the board. A pawn in their game.

What you have to understand is that we, as a society, are the pawns, the human race, the worker bees, the collateral damage, the replaceable. This may sound pretty pessimistic to someone who may not understand any of the topics that will be discussed in this book but you should congratulate yourself for reading this book. You have accepted that perhaps you have a mind that is open enough to have other opinions thrown your way. Instead of discrediting this idea, from the moment it enters your ears and eyes, you said, "Let me hear your argument, show me the facts, perhaps I don't know everything I thought I knew."

What if I told you that flat earth is the single most important concept to wrap your mind around before you truly enter the Nebuchadnezzar and awaken from the Matrix? Shout out, Truth Hooligans! Flat Earth links everything from Tartaria to the great reset, to the Mandela effect, to mind control, from MK Ultra to Manchurian candidates, and on and on and on. Before you begin to understand the game of chess that is being played you must first realize that you and everyone else is a **pawn**, and then you can begin to examine what we live on.

*Let me ask you one question before we get started and I encourage you to take time before you answer this....*

12 Pawns In A Game

Without using NASA or any other "space agency" how do you know we are living on a spinning ball shooting through "outer space?" Can you feel the earth moving? Can you see the curve? Do you feel the wobble? At what point does physical nature step in and you experience vertigo when you travel from the northernmost part of our world to the southernmost? You see it is claimed that those of us who believe in flat earth are unscientific and ridiculous to believe in the archaic teachings that the earth is flat. However, this is simply just a tool used to make this movement look silly and what is the exact opposite of what is in reality. Science isn't a tool that you can believe in, the catchy phrase that was coined during the pandemic of 2020, "trust the science" is moronic at best. As science is used to answer questions, an applied tool to help us discover the unknown. That's like saying "I believe in this hammer because it helps me build things." It's stupid, so stop saying it. The science that we apply is natural sciences. That which is observable, testable, and repeatable. You'll soon discover that those in the mainstream science community rarely use any of the three tools mentioned above to test their theories. Wouldn't it be interesting if Napoleon was right? *"History is a set of lies that people have agreed upon."*

Perhaps we should just take the word of a man who has supposedly been to the moon. …"outer space"...

**"There are great ideas undiscovered, breakthroughs available to those who can remove one of truth's protective layers.." *-Neil Armstrong***

What do you think Neil Armstrong meant by this? I was unaware that truth had protective layers.

The title of the book should be explanatory enough but if you are unsure of what you are getting into, here is a synopsis: the earth is indeed flat and motionless. This book will provide evidence that is both circumstantial and empirical to present more than a theory that the Earth we live on is flat, motionless, and indeed stationary. Some of this may seem hard to digest however, when you start unlearning everything you have been taught it is a much easier pill to swallow. A pill that is laced with nothing but truth and honesty.

*"It is easier to fool someone than it is to convince someone they have been fooled." -Mark Twain*

A quote that remains more relevant today than ever before. We constantly exist in a rat trap that is never-ending with a debt-based money

15 Pawns In A Game

system. Built-in the shadows when good men who never would have allowed such a thing were at their homes for the holidays. Plotted in the shadows, born on an island by the name of Jekyll.

In 2001 we witnessed two planes crash into two towers (or did we?) and fall into their footprint. Explained away by 19 hijackers and the sheer hatred for the American way of life. That day changed the world, it changed all of us, it changed me.

We witnessed marathon runners run past the finish line with their arms raised in the agony of completing such a demanding event when suddenly an explosion wounded dozens and killed a few. Allegedly all perpetrated by the "Muslim extremist," "boogeymen if you were to ask me."

We witnessed "Sandy Hook", "Fast and Furious," war after war after war all in the name of "fighting terrorism." From the outlandish case of

"Kids in cages," to the declassified documents providing evidence that Kennedy was not assassinated by a lone shooter. Hell, it was admitted in a civil court of law that Martin Luther King Jr. was assassinated in cooperation with local, state, and federal law enforcement agencies but you would never learn this in your history books!

Yet the narrative never seems to match the evidence in front of our eyes. Over the past dozen years, we have seen the legalization of false "evidence" provided by the mainstream media, textbooks, and our television screens with no repercussions. Generation after generation has been raised to never question authority and to trust without question, follow without asking, and believe the information that we are given. So in essence pawns…"worker bees."

Our great grandparents, grandparents, parents, children now and then, have grown up this way but as I am writing this book we are seeing a revelation in the likes that humanity has never seen before. A veil of deception is being lifted from the eyes of humanity. A shift in society is taking up strong wind and a resistance to authoritarian rule is beginning to unfold.

---

*"All I want to say is they don't really care*

*about us…"*

*-Micheal Jackson*

20 Pawns In A Game

# CHAPTER TWO

# How I Got Here…

*"It's a big club, and you ain't in it!"*

*-George Carlin*

For starters, I'll tell you my story of how I have arrived at the knowledge that the earth is flat and motionless. Six or so years ago, I was well aware of political and worldwide deception, so being labeled a "conspiracy theorist," was nothing new to me. At this juncture in my life, I endured the term. Side note for anyone reading this who does not know that the term "conspiracy theorist" was created in the late 1960's to label those who questioned the official narrative of President

Kennedy's assassination as, "a big giant crazy person." Just so we are clear, if you didn't believe in the "magic bullet theory" you are the crazy one.

To break this term down even further we can look at the two words individually. "Conspiracy." Which means a group of individuals got together to perform an act of sorts with an implemented plan. "Theory," means it hasn't been proven yet. So in essence we have a group of people getting together to do something harmful (more often than not) that simply hasn't been proven yet. Now, those of us in the truth community, the hooligans out there, feel that enough evidence has been discovered to conclude that nefarious acts have taken place by a group of conspirators. 9/11, Pearl Harbor, the Boston Bombing, Sandy Hook, Parkland, the U.S.S Cole bombing, etc. The elites decided that if they can discredit the character of the source then they can discredit the information gained from the source. (Hegelian Dialectic)

How did I reach the point in my life that I began examining the description of the earth, because flat isn't a shape it is a description. I began to question whether or not the earth was in fact what we are told.

Well, that starts with what seemed to be a regular day at work when my co-workers and I were discussing "conspiracies" when a laborer from another company overheard the conversation and chimed in with, "Have you ever heard of the flat earth theory?" At first, I dismissed this as ridiculous! I felt this idea of the earth being flat is used to create turmoil within the conspiracy world as outlandish claims such as this make all other theories in the community appear to be absurd, and laughable at best. Nevertheless, I thought I would do my due diligence and at least research the topic and thus would be able to put this narrative to rest. By doing this I felt that I could help and add strength to other arguments in the truth community as being factual in reality. We didn't need wild theories muddying the waters.

Yet as I began my research I grew angry. Not because I felt that my time was being wasted but because question after question was being answered and another door was being opened. As I had these questions answered and I walked through the door that had been previously opened, more questions, more answers, more doors, more questions, more answers. This trend continued for roughly three years until I reached a crossroads on this journey which I began just a few years previous. I felt as if I had stumbled upon the mother load of all information concerning the nature of earth which sealed the deal for me and my research. Now I had to make a decision, what do I believe?

Throughout my research, I began looking at the entire argument of the shape of our earth and equating it to a story we are told as children. Sam Tripoli, host of the Tin Foil Hat Podcast has said, "Santa Claus is the first conspiracy theory we all learn and teach our children." I wanted to give a huge shout-out to Sam and his crew for inspiring us to start our show and truly helping us reach our dreams of starting a podcast. And without the supporting advice of Gordon Rochford from Those Conspiracy Guys there would be no WBC universe.

The belief in the heliocentric model is equivalent to that of a child believing in Santa Claus, the true first conspiracy and ironically, the first true conspiracy. As children, we believe that Saint Nicholas is mystical and magical. A jolly fellow that lives in the coldest fucking place on earth with little elves making toys year-round to deliver them in one night. Even though it defies our basic understanding of how the world works it must be true. How could it not be? Our parents, loved ones, families, friends, and people we trust most wouldn't lie to us. Right?

"Mom, Dad, how does Santa Claus fit down the chimney with all those presents?" "How do reindeer fly?" "How does Santa Claus make it to all the children all around the world in one night?" Do you believe in magic?" "magic." From a very young age, we teach our children this idea and we consistently reinforce it year after year until we deem it necessary to tell the children that everything we have told them about Santa Claus is a lie. Most children, I know I did, react with some aggression and pushback. This is called cognitive dissonance.

Webster defines "cognitive dissonance" as: *"psychological conflict resulting from incongruous beliefs and attitudes held simultaneously."* In other words psychological conflict with beliefs that are incompatible with what you have been taught or have learned. When you tell a child for years that something they have been told by their authoritative figure(s) is not true, usually the outcome is at first disbelief that everyone has fooled them. They will actively deny the fact that Santa Claus is not real and that their belief in Santa Claus has been rooted in pure deception.

Notably, this isn't done out of malice in any way by parents, it is a tradition. Now, extrapolate that on a grander scale but with the heliocentric model. When one challenges the belief of the heliocentric model most people who have held this true for a majority of their adolescent and adult lives immediately people will laugh at this, mock with ridicule and disbelief that anyone could be so stupid. Many cannot believe that they have been fooled and that their belief in Santa Claus (heliocentric model) is false. Many of the same aspects of a child's belief in Santa Claus are equivalent to the questions raised when learning about the heliocentric model. How does water stick to a spinning ball? If Australia is on the bottom of a ball, how are there people on the bottom that are "upside down?" "If water always finds its level and the earth is made up of 70% water, how is the earth a ball?" Do you believe in magic? "gravity" a.k.a. "Magic" i.e. Santa Claus sliding down a chimney, flying reindeer, the magical journey of a jolly fat man in a red suit that visits every child in the world in one night. When an idea such as the heliocentric model is challenged most people immediately scoff at this, mainly because the challenge threatens everything they have been taught about the nature of who they are as people. Now this might sound like an exaggeration however, it is much different than telling someone that 9/11 didn't happen exactly the way they were told. It is much different because what this demonstrates is that there are three very simple hurdles in one's mind and it becomes rattling to the core once you start understanding it.

So ask yourself these three complex questions:

*One, have I been lied to since I was a child?*

*Two, if they have lied about what we live on, what else have they lied about?*

*Three, is there a creator?*

If we live on a flat, motionless earth, there must be a creator. If they can lie about the nature of the thing we live on, could they lie about something that seems much less complicated perhaps?

29 Pawns In A Game

Vaccines? Fluoride in drinking water and toothpaste? They promised glyphosate isn't harmful, right? Nah asbestos isn't harmful to your health, right? Weaponized syphilis anyone? Shout out to Tuskegee. People are creatures of habit and lazy when it comes to change and most importantly *research*. If you change the way I think, what I do, say, and believe, if you pull me out of my comfort zone, I will run, kick, scream, put up a fight, and will not accept the ideas with which you present. This is the mindset of a majority of the humans populating this earth. If all of this is true that would mean I would have to go back and re-evaluate everything I thought I knew. Every conversation I have had, every thought I have had, every belief on which I held to be true. This terrifies the common human and I can't say I blame them. I was the same way. This shook me to my core as well, but after three years of having open-minded discussions and doing my independent research, I had to draw a line in the sand. Do we live on a spinning ball shooting endlessly through outer space? A cosmic accident with no real rhyme or reason as to how this all started. Evolved from a giant soup of muck that eventually throughout billions and millions of years became what we could only describe as a perfect accident. Or perhaps just maybe, do we live on what our senses tell us, a flat motionless earth that was created by a higher power? Could it be that we as human beings are unique in every way possible?

This brings me to the point of the people in the world being used as pawns in a game. I hate to break it to you but you are a pawn and that includes myself as well. Critical thinkers are bad for people who want to keep power and for most people it is hard to believe that these people who will willingly lie to you, cheat you, and deceive you are indeed psychopaths. You, like most of us, since we are not psychopaths, can't even fathom what it would be even kind of like this evil. Yet throughout history, we seem to have been taught the idea that Adolf Hitler was a psychopath and now all world leaders, all politicians in power, whether that be in the United States or abroad are not psychopaths. This is not only false but a statistical improbability. The level of deceit is constantly pushed into our faces from the MSM (the mid-level pieces) to television programming, social media, etc. The MSM has recently gotten even worse, due in part to the legalization of false propaganda with the repeal of the Smith-Mundt Act. In 2012 under the guise of the NDAA (National Defense Authorization Act) the powers at be under the Obama administration essentially legalized the use of propaganda on what was originally intended for outside forces or non-American citizens became the focal point of the deep states' plan to influence the minds of all

Americans by using propaganda on United States citizens through the means of controlling the narrative through the televisions, newspapers, the mainstream media and social media.

What does this have to do with the topic of this book? These last two paragraphs lay the foundation of my original statement, that we are pawns in a game. Under this statute that was the NDAA, the government can now lie to its people. Arguably, they have always lied to their people, however, this statute, in particular, made it perfectly legal for the United States government to use news media Outlets, magazines, television programming, etc., to openly use propaganda against its people with no repercussions whatsoever. They can twist the narrative, sway opinions, rally dissent, rationalize hatred for another group, and as stated in the NDAA, under article 114, "none of the information needs to be verifiable as true." The deep state has now legalized lying to the American people to activate the truly evil ideology of the Hegelian Dialectic. Create the problem, observe the reaction (which can be altered in any way they see fit), and provide the solution. We all watched in horror as the world lost its mind in 2020 with the CoronaVirus "pandemic." The "Coronavirus" of 2020 was a textbook Hegelian Dialectic ideology that we have yet to see the outcome of. Judging by the propaganda, I would assume the NWO would like every single human being to be vaccinated and chipped. As the end of 2021 approached we saw countless studies on the damage done to human beings at the expense of this experimental mRNA gene therapy drug. We are now in 2023 and they are admitting they were wrong and asking for a mooligan. However, I don't believe this narrative has finished just yet. If you haven't figured out the game yet, you haven't woken from the Matrix.

34 Pawns In A Game

We have all seen the movie The Matrix. If you haven't, what are you doing? This movie is a damn biography. It is the perfect allegory for what is taking place in the world around us. Narratives manipulated by the controlling media outlets. Politicians bought and paid for. Contrived social issues such as BLM or Antifa. Poison in the food, poison in the sky, poison in the water supply. Until you break your programming and awaken from the matrix you may never see the truth hidden in plain sight. They are lying to you and will continue to lie to you because their Modus Operandi is to enslave Humanity. Remember you and everyone around you is just a pawn, a means to an end.

As Morpheus warned Neo, *"You take the blue pill, the story ends, you wake up in your bed and believe whatever you want to believe. You take the red pill, you stay in Wonderland and I show you how deep the rabbit hole goes. I'm only*

*offering you the truth, nothing more. "*

**Do you still believe in the reality with which it**

**has been presented?**

37 Pawns In A Game

# CHAPTER THREE

# Biblical Evidence

*"To the Chief Musician. A Psalm of David. The heavens declare the glory of God, And the firmament shows His handiwork."* -**Psalms 19:1**

This leads us into the first chapter, where I begin presenting my evidence of the earth being flat, motionless, and stationary. Half of the evidence I provide is circumstantial which at best doesn't prove scientifically what I am discussing. It does however lay the groundwork for what I believe is the nail in the coffin for the debate of the description of the earth. The scientific evidence of the earth being flat is astounding, earth-shattering, mostly hidden and unknown to the people of the world. Those who propagate the lie that we live on a ball, shooting endlessly through "outer space," cannot have this information get out to the masses because it would cripple many scientific fields of study.

**"Knowing is having the power of consent, not knowing, one can not consent...which inevitably leaves you open to the possibility, probability of enslavement."**

**Genesis Chapter 1:1-10, 13-19**

*"In the beginning, God created the heavens and the earth. The earth was without form and void, and the darkness was on the face of the deep. And the Spirit of God was hovering over the face of the waters.*

39 Pawns In A Game

*3 Then God said, "Let there be light" and there was light. And God saw the light, that it was good and God divided the light from the darkness. God called the light Day and the darkness He called Night. So the evening and the morning were the first day.*

*6 Then God said, Let there be a firmament in the midst of the waters, and let it divide the waters from the waters. Thus God made the firmament and divided the waters which were under the firmament from the waters which were above the firmament and it was so. And God called the firmament Heaven. So the evening and morning were the second day.*

*9 Then God said, "Let the waters under the heavens be gathered together into one place, and let*

40 Pawns In A Game

*the dry land appear," and it was so. And God called the dry land Earth, and the gathering together of the waters He called Seas. And God saw that it was good.*

*13 So the evening and the morning were the third day. Then God said, "Let there be lights in the firmament of the heavens to divide the day from the night, and let them be for signs and seasons and days and years, and let them be for lights in the firmament of the heavens to give light on the earth" and it was so.*

*16 Then God made two great lights: the greater light to rule the day, and the lesser light to rule the night. He made the stars also. God set them in the firmament of the heavens to give light on the earth*

41 Pawns In A Game

*and to rule over the day and over the night and to divide the light from the darkness. And God saw that it was good. So the evening and the morning were the fourth day.*

Now if you aren't a religious person the above verses won't mean a whole lot however I have heard numerous accounts of people, "finding God " after researching the shape of the earth. If this doesn't apply to you that is perfectly fine and you can skip down a bit to the next section.

---

However, if you are a Christian, believer, or a follower of Christ you may want to re-read the first chapter of Genesis again and ask yourself if you truly believe in what the Bible says. You CANNOT pick and choose the verses you want to

42 Pawns In A Game

believe and the ones you don't. If you read this chapter and believe that it does not teach a flat earth then I encourage you to re-read again and then ask yourself this unsettling question. "Why does the Bible, the book that I believe to be a guideline for my life, not match with what I have been told and taught my entire life through curriculum, movies, television, or even my pastor?" Why is it that the story of creation systemically leaves out information that we have been told about the earth? Why does the story of creation say that God created light and then divided darkness from light? Shouldn't the Sun have been the created light, which wasn't created until verse 16 ("a greater light to rule the day") which was the fourth day? What was the Earth-orbiting for three whole days before the creation of the Sun? Why in verse six does it state that

*"A firmament in the midst of the waters, and let it divide the waters from the waters. Thus God made the firmament and divided the waters which were under the firmament from the waters which were above the firmament."*

This implies that there is water above the earth's sky which would mean there has to be a physical separation by a firmament. Why does the Creator make the distinction of water being above the earth in the heavens? This fits the description of the Swiss physicist Auguste Piccard who had set a record in 1931 reaching the highest altitude ever (before the creation of NASA). Piccard stated that the earth appeared to be a *"flat surface with upturned edges"* and described the medium in which his high altitude capsule was traveling appeared to be that of the consistency of water. Why does the Creator conveniently leave out the creation of planets? Why does the Creator leave out the part where he decided to rotate the earth at 1,000 mph and send the earth into orbit around the sun at 66,000 mph?

You could say that when the Bible was written humans didn't know about these other wandering bodies and perhaps this is a sufficient answer to this question but you must answer all of the other questions which were proposed in the paragraph above. Perhaps we can take a look at some other verses and ask even more questions about the Biblical evidence concerning the truth of the nature of our earth.

**Joshua 10:12-14** states, *"Then Joshua spoke to the Lord in the day when the Lord delivered up the Amorites before the children of Israel, and he said in the sight of Israel: "Sun, stand still over Gibeon and Moon in the Valley of*

*Aijalon."* **13,** *"So the sun stood still. And the moon stopped. Till the people had revenge upon their enemies. So the sun stood and the moon stopped. Till the people had revenge upon their enemies. Is this not written in the Book of Jasher? So the sun stood still amid heaven and did not hasten to go down for about a whole day.* **14** *And there has been no day like that, before it or after it, that the Lord heeded the voice of a man; for the Lord fought for Israel.*

If you are a believer do you think this is metaphorically written? Do you think that God made the sun and moon stand still but he meant the earth because according to the heliocentric model the sun doesn't move the earth does?

How about **Job 40:22**? *"It is He who sits above the circle of the earth, and its inhabitants are like grasshoppers, who stretches out the heavens* *like a curtain, and spreads them out like a tent to dwell in."*

Perhaps the Creator didn't know that a circle is a 2-dimensional shape and He is sitting above a globe or oblate spheroid as Neil Degrasse Tyson likes to call it. Are those curtains about the firmament mentioned in Genesis, the physical barrier between the waters above and the waters below? What is that old saying from the satanic bible? ***"As above, so below."*** Shit even Satanists know what the deal is!

**Psalms 104:5** *"He established the earth upon its foundations, So that it will not totter forever and ever."* **Psalms 93:1** *"The Lord reigns, He is clothed with majesty; The Lord has clothed and girded Himself with strength; Indeed, the world is firmly established, it will not be moved."* **Psalms 96:10** *"Say among the nations, "The Lord reigns; Indeed, the world is firmly established, it will not be* *moved; He will judge the peoples with equity."* One of my personal favorites; **Job 38:14** *"The earth takes shape like clay under a seal; its features stand out like those of a garment."*

What does **Revelations 7:1** say about the description of the earth? *"After these things, I saw four angels standing at the four corners of the earth holding the four winds of the earth that the wind should not blow on the earth on the sea or any tree."* Corners of a globe huh? Perhaps the four corners are about the edges of the flat earth? I don't assume an edge but perhaps. You must ask yourself if you believe in what is written in the Bible, how do you interpret this? I could give you about 150 more verses but I think at this point you are getting the picture.

If you are a believer in Christ and what the Bible teaches you cannot brush this over as something that you choose not to question because

48 Pawns In A Game

it goes against what you have been indoctrinated to believe. You have been lied to your whole life and I understand I am shaking your entire belief system to its core but you must realize that I am not the one who has been lying to all mankind for generations. I am simply the one trying to break the cognitive dissonance.

---

Deception in the church is nothing new. We are all too familiar with the Catholic Church sex scandals. Mega pastors lock their congregation out during a natural disaster. The Christian churches and all other religions are not exempt from this corruption either. I believe that the Christian church has folded its knee to the outright corrupt doctrine. At large I believe they teach lessons that are non-biblical and do not follow any evidence presented

49 Pawns In A Game

within the Bible to collaborate their teachings. Why isn't the Book of Enoch in the Bible? In my belief that book was removed from the Bible to hide evidence of the ancient bloodlines of the Illuminati or the "Nephilim."

---

## The Book Of Enoch

A book could be written about the Book of Enoch and why it is so controversial. I do believe that this book was purposefully taken out of the Bible because it discusses some taboo topics, to say the least. Hidden ancient bloodlines anyone? The Book of Enoch is an interesting read, to say the least. The ending of Enoch is even more interesting. The only person mentioned in the Bible who did not leave this earth the same way we all do. Enoch did not die. In **Genesis 5:24** it states that *"And Enoch* *walked with God: and he was not; for God took him.* " Other translations say that Enoch walked faithfully with God and God took him away because God was pleased by Enoch. I have asked the question many times of pastors and teachers of different churches, "Why isn't this, what would seem an extremely important book, in the Bible?" Most of the answers remain the same, "the book of Enoch wasn't ordained by God Himself, therefore it didn't belong with the prophetic books of the Bible."

Now this isn't any of their faults for not knowing why truly, they have been taught the same things we all have been taught. However the more you read and research the more you will be amazed with the knowledge you will gain. I believe the Book of Enoch was taken out of the Bible on purpose, during the council of Naicia. Reasons for this may sound far-fetched however they paint an even more in-depth picture as to who the movers and shakers of this world are. The Book of Enoch lays out a systematic timeline from the personal account of Enoch, a man so holy and pleasing to God he was able to walk alongside the creator of this universe. A man so Holy and righteous in God's eyes, God decided not to let Enoch fall to death. Even so, the book is not denounced by any religious scholar I have ever met.

The Book of Enoch walks through a more detailed account of how the Nephilim were created, why the Nephilim were on the earth, who they were, and how giants were created. More history the chess masters want to hide from mankind.

The book also details what exactly we are living on. Take **Chapter 18:2** for example: *"I surveyed the stone, which supports the corners of the earth."* Ah back to the oblate spheroid having corners again, huh? About **Genesis Chapter 1:16,** the Book of Enoch seems to piggyback off of this verse by stating in **Chapter 60,** *"How the portals of the winds are reckoned, each according to the power of the wind, and the power of the lights of the moon."* So the moon is its power source or could we say light source? Didn't the Creator say He created a lesser light to rule the night? **Chapter 77:3** says, *"These are the two great luminaries; their circumference is like the circumference of the heaven, and the size of the circumference of both is alike."* So now we have measurements of the two great luminaries being declared by a man who was so holy to God he was taken away and did not die. Is it starting to make sense as to why this book was taken out of the Bible?

Genesis made a very clear distinction of the fact that God separated the waters from the waters. One great body of water on earth and waters above the heavens divided by the firmament. In the Book of Enoch **Chapter 50** says, *"And in those days shall punishment come from the Lord of Spirits, and He will open all the chambers of waters which are above heaven, and the fountains which are beneath the earth. And all the waters shall be joined with the waters: that which is above heaven is masculine, and the water which is beneath the earth is feminine. And they shall destroy all who dwell on the earth and those who dwell under the ends of heaven."* Auguste Piccard comes to mind again...Also another story of a worldwide flood.

The Book of Enoch also discusses the specific DNA code of the "fallen ones." The "watchers," the Nephilim bloodline that still flows within human veins today. If you seem to be having a hard time believing what I just said about fallen angel bloodlines you only need to look no further than the (RH) compound in human blood. The Rhesus factor is a type of protein found on the outside of red blood cells and this compound is inherited from your parents. Scientists claim that your RH factor doesn't affect your overall health however if you are a female and you are RH negative and become pregnant you must have the medical intervention of a drug by the name of Rogam to see your pregnancy to full-term if you have mated with an opposing blood type. If a woman doesn't receive the shot, the RH-negative compound will create antibodies and will attack the proteins of the baby if the baby's blood type is positive. Only about 15% of the human population are Rhesus Negative. The rarest blood type is RH-Null. It is said that only about 60 humans have RH-Null, referred to as "Golden blood." In my belief the RH-Null blood, the rarest of the rare, is the closest thing to the Nephilim bloodline, and the second closest is the RH-Negative blood type. My wife being one of those humans whose blood flows with that of the Nephilim, (she hates it when I make that joke) is also O-. Which makes her a universal donor. I always make jokes with her about her damn angel blood and how her health issues are related to the fact that her fallen angel blood isn't meant for mixing with things of this earth, but that's a whole different story. In short, the Book of Enoch discusses two major taboo topics in the scientific community. It teaches a flat, motionless earth created by God, and it discusses the fallen angel bloodline of the Nephilim and the offspring of the Nephilim; earthly giants. This history is hidden on purpose because it tells a completely different story than that which is presented today. Think of it this way, if giants were real and walked the earth, then the Bible is correct in its claim of giants existing, then what else is correct from the Bible? See how this presents an issue from the heliocentric cult? In the movie Clash of Titans, giants do battle with each other on screen demonstrating their god-like powers and strength. Was this just a movie or was this a story being told as a piece of forgotten history? I challenge you to do your research on whether or not giants existed but I would start with the Kandahar Giant.

---

Even the story of the rapture is questionable in my opinion as there is no evidence of this event coming to fruition in the Bible and the word, "rapture" is not even once mentioned in the Bible. We have to be very careful with how we read texts with which we hold to the highest level of truth. If we claim that we are members of the truth community then we cannot accept what we are told at face value. The deception at this stage of the game is too obvious. The trouble here is how do we lift the veil of corruption and deceit that has shrouded the truth for so long? How do we break through the narrative that is controlled by the most powerful corporations and figureheads in the world? How do we not even realize the organizations that we should trust the most, like the fundamental religious bodies are also telling lies through a corrupt past?

Why is the mainstream media so quick to take control of a story?" Immediately issuing a one-way street of information and any side streets of that road are deemed a conspiracy. Why are so many news outlets, social media platforms, and television programs quick to claim that any alternative sources of information that go against the status quo, are "debunked?"

59 Pawns In A Game

60 Pawns In A Game

# CHAPTER FOUR

# Circumstantial Evidence: An Overview

*"...if the American people ever find out what we have done, they will chase us down the street and lynch us."* -George H. W. Bush

Before we get into the true description of the earth, we must go through some rabbit holes of evidence that is blatantly being hidden from humanity. I'd like to go through the things that flat earthers (for lack of a better term) do not believe. Or at least I don't. As a community, you could imagine that there are many different theories within theories, which is perfectly fine but we have to be careful when we are deciphering information so as not to be fooled by **misinformation**.

*"The earth is a flying upward pancake because gravity is real and this is how this force is seen in our world."* A perfect example of where not to do research is the *"Flat Earth Society."* This is bad information and controlled opposition, their job is to purposefully give out faulty information that will be used to discredit the entire movement. For example, the FES claims that while the earth is flat and not rotating, the disk-shaped object that we inhabit is currently shooting directly upwards in space which would give us the downward force (not a force acting like a force) of gravity. This is of course idiotic. Recently they have written articles

claiming that perhaps the earth isn't flat, it is in fact "donut-shaped." This is done on purpose. If you were to know nothing of flat earth and went into let us say, Google to search flat earth this will probably be one of the top search results. This is meant to lead people to this ridiculous information so that once they see this they will laugh, discredit the movement, and file that claim under, **"case closed and do not open until next Christmas."**

If you remember, President Barack Obama mentioned the Flat Earth Society in one of his speeches during his presidency. **"We do not have time for a meeting of the Flat Earth Society."** Why would the sitting president of the United States mention the Flat Earth Society at a press conference concerning climate change? Do You believe this was done by accident? Of course, it wasn't because you should know that nothing happens by accident. One of the most popular presidents in the history of the United States, mentions the Flat Earth Society, and people flock to search the term Flat Earth Society and read articles about floating donuts and rising space disks and people make a run for the hills. During the second term of Barack Obama, the movement of questioning the true nature of the earth was growing in popularity. It was the third most searched term next to Donald Trump and Hillary Clinton. So the campaign to squash the movement was launched and active within social media sharing platforms such as YouTube, Instagram, Twitter, and Google.

**"We do not have time for a meeting of the Flat Earth Society..."**

**-Barack Obama**

That was all it took for enough people to scratch their heads. Imagine the curiosity of the masses watching their beloved president mention something so foreign to them. Saying to themselves, "I have never heard of the Flat Earth Society!" Systematic, purposeful shutdown of a theory in which the elites do not want their "pawns" to be discovered as this would be the first step to mass awakening. The elites would begin to lose their power, almost overnight. Could you imagine if a spokesperson for NASA came out tomorrow and said,

*"... we have been lying to you for years and years. Not just our organization, but all world space organizations, as well as scientists, school teachers, and your government. Through no fault of their mind you, there are only a handful of us that know the truth. We do not live on a spinning ball shooting through "outer space." No actually, quite the opposite, we live on a flat plane with an enclosed dome above us and we were put here on*

*purpose by an all-powerful creator. We might want to figure out what this creator wants from us. No, unfortunately, we will not be taking questions on this matter.*" Like an episode of Dave Chappelle, heads would explode, chaos would ensue, entire monetary systems would collapse and the cast of the Ancient Aliens T.V. show would be pissed.

The FES is nothing more than a controlled opposition spreading disinformation. As we move forward I encourage everyone to focus on where they are getting their information from especially when doing further research into this topic. The United States government has made no secret about their wrongdoings when involved with other countries. Whether it be elections, war, or societal issues, our government here in the United States is more than likely involved. Hell! We have been at war every year that I have been alive except the four years Donald Trump was president. It baffles me that people ignore the fact that The United States government has lied to its people. Do you remember the Tuskegee experiments? Injecting weaponized syphilis into unsuspecting black people leaving many of them dead. How about Agent Orange? Fluoride, smoking, smoking indoors, smoking while pregnant, asbestos, The Gardasil shot for HPV, Benzagzi, and the Gulf of Tonkin incident. We could go on and on…

The Gardasil shot for HPV spread widely during the 90s, literally crippled hundreds of thousands of women. How quickly do people forget they are scared into believing another virus is out to get them? Perhaps…everyone who was injured by that shot is either dead or too incapacitated by the side effects of the shot to notice. Since vaccine manufacturers are exempt from vaccine injury cases, that shot is making a comeback! Do you want some real-world advice from a first-hand account of HPV? My wife and I are both HPV-positive and have never once experienced a single symptom related to HPV. Maybe because we haven't been treated for it? Shout out to Magic Johnson, what a walking miracle he is! Statistically speaking, of course, the sooner you are diagnosed, the sooner you start treatment, the sooner you die.

There are so many examples I could give and yet there is no possible way the government would ever lie to us about the nature of the thing that we live on. Right? I mean being wrong 99 times out of 100 doesn't mean they are wrong about the last one. One very interesting argument is that there is no way that a conspiracy of this magnitude, this size, could be this widespread. Again the CIA created the word "conspiracy theorist" to make people look crazy. Well look at it this way, in 2020 a very small group of people convinced the entire world, and I mean the entire world, to shut down for the big bad "Coronavirus." " *Stay in your house, lock the doors, don't hug grandma! 6 feet apart you murderer! Mask it or casket it!* " (*That one is my favorite*) This virus was/is so deadly that it kills less than the flu, is curable, (and I use this term loosely) with natural remedies, and in my belief, is non-transmittable. I truly do believe in terrain theory and encourage anyone who has made it this far in the book to research this topic before you ever head back to the doctor. Germ Theory is total hogwash and this one goes deep but this is another topic for another day.

Entire economies were flushed down the toilet, lives ruined, human beings turned against one another, people afraid to leave their houses, wearing diapers on the face while driving alone in their cars. I bet there is someone right now to this day in the shower alone wearing a mask. How many people need to be organized to collaborate on this lie? 10? 20? 100? It isn't that many when you think about it. It was a perfect scheme, like yelling fire in a crowded theater, letting chaos do the rest. Yet again we meet the Hegelian Dialectic.

As we travel through this book I will lay the groundwork for what are the key points that sealed the deal for me in my journey of searching for the true nature of our earth. This may not seal the deal for you, however after reading this book in its entirety and then fact-checking it for yourself, (I mean fact-checking this information, not just looking up a Snopes article) you will at the very least want to keep digging. I will present multiple key points of what I would consider circumstantial evidence which doesn't necessarily prove anything, yet it does demonstrate strange anomalies and raise definite eyebrows at things in which you may have not known. Things that we may have originally taken for granted as 100% factual. The remaining evidence that I will present will be the nail in the coffin, hardcore facts, and undeniable scientific experiments that no "scientist" thus far has been able to deny. Let me be clear, you will hear many times that "flat earthers" are **"science deniers."** Globers claim that we do not believe in science. What they mean by this is we don't believe in pseudoscience. A lot of people believe that if it is written in a textbook, it must be true. Throughout this book, evidence will be presented proving that many scientists no longer actively use the scientific method. If you recall, back in middle school, junior high, and high school, the scientific method was the first system we should use when approaching a scientific question.

I have had countless conversations with numerous people about this very topic and the conversations usually end up in the same place. Deadlock with the look of anger on their faces. Or sometimes, keyboard warriors hurling insults. Some of the more comical responses are "Umm you just have to google the earth is round, hello!" or "Wow, you believe the earth is flat, how stupid can you be? We have known the earth is a spinning ball for hundreds upon thousands of years." Or my particular favorite, "You do not believe in science!" This is usually after I ask for proof of what they believe and it comes down to those who were once just calling me a "science denier" now saying, "Just google it," or "Do your research I bet you won't." I find it very interesting that if I were to ask you what evidence you have in your mind right now as to why you believe the Earth is a spinning ball shooting through space, I would bet that most of you would say, "I just know it is, because I have been told this my whole life." Ah and there it is, I have been told this so it must be true. There is no way my authority figure would lie to me, would they? Interesting question isn't it?

Understanding science I think is one of the most important aspects in understanding facts versus believing in fiction. Science helps answer questions that arrive in our world when something that is naturally recurring is unexplainable. I believe that if common folk could apply this methodology to their very own daily lives, less illusion would be achievable by our handlers. I will do my best throughout the remainder of the book to demonstrate and break down as simply as I can my understanding of how the Earth is without a doubt, a flat and motionless plane using the scientific method. Easily demonstrating how the heliocentric model doesn't hold weight when actual science is applied to it.

---

# "To Deceive By Lifting Up"

*"A Nazi, a Science Fiction writer, an Occultist, and a Racist Cartoonist walk into a bar…"*

What do you get? NASA! That's not a joke either. The creation of NASA is credited to four individuals that you rarely hear about who created the establishment that we know today as NASA. This chapter will be devoted to these individuals and how they are not the first people you would think would be the founding fathers of NASA.

Verner Von Braun is credited with being the father of the Rocket program in the United States and I say this reluctantly, "helped put a man on the moon in 1969." Walt Disney, a known occultist and oddly racist man who created cartoons for children, decided that he should also play a hand in putting men on the moon. Jack Parsons, an occultist, sex magician, scientist, and absolute lunatic who had a very dark ending involved himself in rocketry at a very young age and tossed his name in the hat to help NASA get its wheels turning. Lastly, but certainly not least, L. Ron Hubbard, (you've got to be kidding me), you know the father of Scientology, you know the cult that believes they will fly a pirate ship into outer space or some shit...yeah those are the founding fathers of NASA. Now I could expand on them greatly to add the silliness of this but I will leave that to the revised edition of this book.

I believe that NASA is a Satanic organization. Your belief in Satan is irrelevant in this argument because these men believed in the father of lies. Now before you laugh and walk away, know that three of the members are openly admitted Satan worshipers and occultists members. Is it out of the realm of possibility that these people wouldn't incorporate their ideologies into their field and craft? We now know how sick Walt Disney was and the allegations against the Disney organization with not only child pedophilia and the disappearance of children but their involvement with high-level CIA mind-altering experiments such as MK Ultra, as well as Project Monarch. It is an absolute fact that Walt Disney was funded by the CIA when purchasing land in Florida to build

"Disney World." Verner Von Braun was a Nazi scientist for Christ's sake! Are we to believe when Braun was brought over under "Operation Paperclip," with hundreds of other scientists that he had a change of heart? Not to get off on a tangent but the whole "Nuremberg trial" was an absolute clown show for the world to believe that Nazism was dead. The Third Reich didn't lose World War 2, Germany did. How many people were hanged during the trial? Like 10? It is well known that the Nazis were obsessed with the occult and reaching a higher sense of self-understanding by attempting to reach a higher human state on a metaphysical level.

The word NASA in Latin means, "to lift" or if pronounced NASA' it means, "to deceive" combine those two words in any fashion you like and you either get to lift to deceive or to deceive by lifting. Interesting isn't it, what do the words mean?

This might all sound ridiculous and there is no way any of this could mean anything other than sheer coincidence. Well is it any more crazy that in the 2000 presidential election, we had two choices for president, republican nominee George W. Bush Junior and the democratic nominee by the name of John Kerry? Did you know that both candidates went to Yale University? Well that isn't that strange most of the politicians in Washington attended an elitist school such as Yale, Harvard, or Princeton. What is strange is how both candidates belonged to a club, a secret club, that only accepts 15 members a year. Meaning that there are only about 800 or so living members at any given time. If we did a little math on that, the 2000 census puts the US population around 281,710,913 so just under 300 million which is roughly our current population in 2021. Would the percent chance if we took (for the sake of it) all 800 living skulls and Bonesman, that both candidates would belong to the same secret society out of the entire population of the United States be 0.000003%? (Of course, we can't take the entire number of the population for that equation because not every citizen could be president, etc.) But I'll put it to you like this...it's a slim fucking shot. I'm sure that is just a coincidence too, right? Once again the illusion of choice and freedom is presented every four years, when in reality pawns are used and the masters play the game of chess.

Every "planet" is named after a false Demi-god which in my belief is done on purpose as the heliocentric model as a whole is a satanic cult. Saturn is the god of generation, dissolution, plenty, wealth, agriculture, periodic renewal, and liberation. Damn! Saturn has a full plate, Jupiter is the god of the sky and thunder, Neptune is the god of the sea, Mars is the god of war, Uranus is the god of the sky Pluto is the god of the underworld (nice) and so on.

Now this isn't proof of a flat earth mind you this is more or less evidence that I am presenting for NASA being a cult, a satanic cult that worships the sun, moon, and stars. I'll go a little deeper as far as these anomalies go with NASA and the belief in Lucifer whom they have chosen to worship and hide their doings within the organization itself.

Scientists have claimed that the sun is 93 million miles away. Hmm, I wonder how they got that number? This number varies from time to time but it remains close to 93 million miles away with an amazing size of 40 times greater than the size of the Earth and has been classified as a super dwarf. With the sun being 40 times greater than the size of the Earth science would have some explaining to do, no? Seeing as how the sun and the moon appear to be the same size while they circumnavigate the sky above us. Well, you see even though the sun is this large at a distance of roughly 93 million miles away from the earth the sun is also 40 times the distance from the earth. So the sun is said to be 40 times the distance and 40 times the size which is why the sun would appear the same size as the moon no matter its location in the sky. Now believing in a creator sounds ridiculous because space is a violation of natural law which we will get into this later but how amazingly coincidental that the universe (which remember was created from an explosion of material that didn't exist and expanded over billions of years to give us of what we have now) gave scientist such luck that due to the sun's relative size and distance would solve its question. However, let's look at this from a nonsensical perspective, shall we? How do scientists know the distance to the sun? Do we know what the sun is made of? The answer to both of those questions is no. No tool is accurate enough to measure the distance to the sun and there are so many variables when doing these measurements that most of the quantities that are "measured" are assumed. The same goes for the question of what the sun is made out of which would be important to know the answer to before asking how large it is and how far away it is, that goes for the moon as well.

Everything we have been told about the moon is a baseless claim with no empirical evidence, always falling just short of the scientific method. As we continue this journey, I encourage you to think about the evidence that I am presenting concerning actual science and not pseudoscience. Apply these theories and hypotheses using the scientific method and you will be shocked at how little the "science community" actually uses it!

According to the heliocentric model the earth is supposedly rotating at 1,000 mph at the equator and is orbiting the sun at roughly 66,600 mph. The solar system is orbiting the sun at roughly 666,000 mph and the Milky Way galaxy in its entirety is shooting through "outer space" at roughly a million miles an hour. None of these motions are visible or sensible to humans here on earth and mind you according to scientists like Neil deGrasse Tyson, the earth is wobbling as it spins. The earth is also claimed to be on an axis of 23.4 degrees which from a 90-degree angle is 66.6 degrees. Again this isn't evidence of the earth being flat, however, I did make a rather outlandish claim that NASA itself was a satanic cult organization with 3 of its 4 founding members being Satanists and occultists themselves. With that being said, do you see the odd pattern here? 66,600 mph, 666,000 mph, and an axis of 23.4 off an angle of 90 degrees leaving 66.6? Odd pattern isn't it? Again this paragraph is more or less circumstantial evidence however it is directly relevant to what we are discussing in this book as the people who control our societies and make decisions in this world, the "movers and the shakers" if you will, are psychopaths. It only gets more strange from here.

NASA was developed in 1958 under Dwight D. Eisenhower and in 1961 a very peculiar treaty was signed concerning Antarctica. A treaty that was originally signed by 12 countries which now consists of 54 parties and will not be negotiated until 2050. This treaty set very specific guidelines that no country shall ever lay claim to this area for research purposes or military advancement. This is very odd considering countries fight over territory all the time. How weird is it then that a majority of the developed countries in the world agree to never go there for any research purposes? This is peculiar considering the timing of the treaty and after a particular television interview was conducted by Admiral Byrd who was the youngest admiral in the United States Navy. According to Admiral Byrd, "The land beyond Antarctica is the most important piece of the world left for science." Beyond Antarctica? What does this mean? Operation High Jump lasted from 1946 to 1947 and consisted of 4,700 men, 13 ships, and 33 aircraft (33 aircraft huh, Freemasonry numerology) Dozens of expeditions were conducted exploring the Arctic ice shelf, but as you can imagine after this interview they scrapped the project and this was Admiral Byrd's last trip to Antarctica. One would have to wonder why after a United States Admiral states that there is land beyond Antarctica that is the most important piece of land left for science, Antarctica would be closed off for universally every country, and with 54 parties now involved in the treaty, we could say that almost every country agrees, Antarctica is off limits. Interesting to note that a section of the Encyclopedia Americana 1958 edition stated very clearly that a "dome" had been discovered. (Concerning Antarctica) "...these flights inland seem to prove featureless in character, talking about a dome reaching 13,000 feet high at about latitude 80 degrees south longitude 90 degrees east." All maps in the encyclopedia up to 1958 had maps depicting a flat earth with Antarctica circling the earth. The same year NASA was developed and three years later the Antarctic treaty was signed.

With the development of NASA in 1958 the government cannot have an Encyclopedia out there stating that the earth is flat, with Antarctica surrounding the earth and an approximation of where the edge of a dome could be. After NASA was developed the encyclopedias containing this information and all information containing research regarding Antarctica and what is truly out there now only come from the extensively trustworthy agency called NASA. Let me present a very interesting timeline for you that will begin to paint a very different picture than what has been presented to the masses.

From 1946-1947 Admiral Byrd had been coordinating Operation High Jump where a series of explorations were conducted in Antarctica to determine how much land was still available there and if certain equipment worked in frigid environments such as the Arctic. Admiral Byrd then appeared on the "Longines Chronoscope" television show in 1954, where he declared that "the land beyond Antarctica is the most important piece of the world left for science." He claims that there are enough resources there, natural resources that all of mankind could utilize worldwide for many years to come. Admiral Byrd dies in 1957 and in 1958 NASA is created under Dwight D. Eisenhower of course with the credit going to Walt Disney, (nazi sympathizer) Verner Von Braun (former nazi scientist, father of the rocket program, incumbent of operation paperclip) Jake Parsons (occultist, Satan worshiper) and L. Ron Hubbard (occultist, science fiction writer). In 1961 the Antarctic treaty was signed and now what has been said to be the most important piece of land left on earth for science is off limits, to all countries' governments for military and or research purposes. In 1962 Operation Dominick took place where a series of high-altitude "nuclear strikes,'" took place and within this operation, Project Fishbowl was set in motion. Sidebar here, "Dominick" means, "of the Lord." If we do some deductive reasoning, and we put two and two together we get, a "fishbowl of the Lord."

Hmm Fishbowl of the Lord? A brief breakdown of project Dominick goes as follows; the United States government conducted a series of missile strikes, "against something" in the upper atmosphere. They were firing missiles into or at something at which point they were detonating at high altitude. Shortly after these tests concluded NASA was formed. Could it be after the discovery of the land beyond Antarctica according to Admiral Byrd and the dome that is mentioned in the 1958 Encyclopedia Americana at specific latitude and longitude they were perhaps trying to see how high the dome went and maybe if it could be penetrated?

---

90 Pawns In A Game

# CHAPTER FIVE

# The Evidence

*"I believe in evidence. I believe in observation, measurement, and reasoning, confirmed by independent observers. I'll believe anything, no matter how wild and ridiculous, if there is evidence for it. The wilder and more ridiculous something is, however, the firmer and more solid the evidence will have to be."*

*-Isaac Asimov*

In the next three sections, I will be presenting the evidence that has sealed the deal on the debate for myself and many others as to the earth being flat and stationary. This is the part where you need to open up your mind. These examples are not wild-the-wall explanations and we will walk through these together. This evidence is ideas that people have probably already thought about throughout their lives but dismissed. "Oh my science teacher knows better than me, so I guess I'll just take their word for it." This is where we must all, including myself, check our cognitive dissonance. You need to open your mind to what you already know about science and how to apply it to the topic in question. Is the Earth a spinning ball, shooting endlessly and aimlessly throughout outer space? Is everything that I have been taught in school about the place we call Earth true? Was this done on purpose or is this just a lack of understanding? Will I be able to accept the answers that are given if they produce results that I am not accustomed to believing in or have been taught to believe?

You see, the masters of the chess game are very good at what they do, they have been playing this game for decades. It isn't as if they just started the other day and decided they were going to be deceiving all mankind in a couple of short weeks. No, they have been doing this since the beginning of time; this is a generational psychological operation. Have you ever heard of M.K. Ultra? A real-life psychological operation conducted during the late '60s and early '70s by the CIA. A real mind control program. But was this the earliest they had experimented with this technology? Would it at all shock you if I told you that Whitey Bulger was a victim of CIA mind control? In 1956 Whitey told an inmate while he was serving his first prison sentence, that he was a victim, a product, of the CIA mind control operation M.K. Ultra. Perhaps, these tests were not only orchestrated during the late 60's and 70's. Did they ever stop? This all sounds crazy enough in and of itself but most people don't understand the fact these puppet masters are psychopaths, we as regular old human beings can't comprehend this since we aren't insane. Most people don't even know certain FACTS that would otherwise make them green in the face and weak in the knees. For example, did you know that Henry Ford was a nazi sympathizer during WWII? He was an adamant supporter of Adolf Hitler and wanted to help create a series of trucks to help the advancement of the German war machine. J.P. Morgan is another great example of a power-hungry psychopath who funded the nazis and made millions off the war, which is why even today the symbol used for the Chase bank is a swastika.

Don't believe me? See for yourself... (just Google it).

Wonder what would happen if we put a cross symbol right in the middle of the Chase symbol? We could go on and on about how these people play games with their "slaves" (you and I) but let's get into the hardcore facts of what changed my mind completely about the earth on which we reside.

*"Outer Space" isn't so spacious*

*-Big-Country*

"Outer space" is a violation of a natural law called the 2nd law of thermodynamics. In short, this law can be summed up as "rate of decay." Over time everything is in a state of decay and is proportional to time. Nathan Oakley gave this example: *"If you painted a room perfectly and did nothing else to it, left it the way it was for years, and came back the room would be in a state of decay.* " Paint could be peeling, chipping, and fading, a smell of must could be floating in the air, and layers of dust would be covering the remains of what was once a beautiful creation is no longer beautiful and a display of neglected chaos. Gas law seems to be something that most "scientists" seem to be okay with violating when they explain the earth's atmosphere, even though it is a naturally occurring law. The earth has gas pressure and space is claimed to be a vacuum (not a vacuum that sucks) just a space (volume of emptiness) having no elements and having infinite space, a volume that is expanding. Due to the naturally occurring phenomenon of how gas operates, the gas we breathe on Earth will eventually fill the available container (volume) of space. In essence, the gas we breathe would escape instantaneously into "outer space" to fill the available space. Numerous times I have brought this very point up during a debate and the common refutation is, "Well there is a pressure gradient." A pressure gradient defined in Webster's Dictionary is *"the space rate of variation of pressure in a given direction.* So the higher we go the less pressure there is. This is true however the evidence of a pressure gradient only demonstrates one thing. That we are breathing pressurized air. This means we have a pressurized system directly next to an unpressurized system with no physical barrier. This is not only a physical impossibility based on all testable observations on earth but there is no experiment that we can do on earth to replicate this environment. In other words, we cannot in any way pressurize gas on earth with the gas being directly next to a non-pressurized system without the gas rapidly reaching its equilibrium. You cannot pressurize gas without a container and this demonstrates the first problem with the heliocentric model. Of course, they like to mention gravity but we will get to that a bit later.

Every earth test that can be done following the scientific method proves that gas cannot be pressurized without a container, volume to be filled and walls to hold in the gas that is pressurized. This is one of the biggest pieces of evidence to support a flat motionless earth that perhaps is contained by a dome. I certainly don't necessarily claim there is a dome, which I cannot prove or test myself, therefore I cannot use this as a shred of evidence, however, the gas pressure is an issue for the heliocentric model. Even with the claim that is regurgitated nonstop in the "scientific community" that gravity holds in the pressure of the gas we breathe on Earth, it cannot be tested. If I remember correctly, appropriately applying the scientific method is a requirement when stating a hypothesis. So why can't we test this theory of gravity holding in gas pressure next to a vacuum? What I do find interesting is the lack of science that goes into what people believe concerning the heliocentric model. The ignorance of gas law is a perfect example of what I and many others call "scientism." Let's continue to walk down this path of gas pressure being an issue. Gas expands in all directions, the only way gas is pressurized is by having the required antecedent to gas pressure, a container. Gas must push off of a container to be pressurized. This is naturally occurring which is why it is considered a natural law. Now "outer space" is a vacuum, an empty container that has no elements and has an infinite volume to fill, or so we are told. So what we have then is an empty container surrounding the earth with no barrier in between the gas that is pressurized on earth and the absence of gas in space which would violate not only the gas law since gas is omnidirectional but the 2nd law of thermodynamics as well. Gas does not escape into space according to the heliocentric model, which as we know from actual applied science would happen instantaneously. If gas is pressurized on Earth, which it is, it would want to fill the available volume of space because that is what gas does naturally. This would happen immediately because the earth is supposedly suspended in outer space and surrounded by the emptiness, the volume, that the gas would want to fill. Remember, the antecedent to gas pressure is a container. Gas wants to fill volume, that volume in "outer space" according to the heliocentric model, it just magically doesn't. *Gravity* is the mystical, magical, unexplainable force that isn't a force that acts like the force that is used to explain away any questions that are raised in defiance of their space ball belief.

Kind of like when your parents say, "Because." In numerous debates that I have been involved in about the shape of the earth, gravity is brought up many times as a way of trying to avoid a question that is unanswerable unless gravity exists. An "end all be all" if you will. What keeps gas pressurized on the earth's surface if gas expands in all directions and wants to fill the available volume which is outer space? "Gravity!" After I wiped away my tears of laughter, I put the final nail in the coffin of this ridiculous argument.

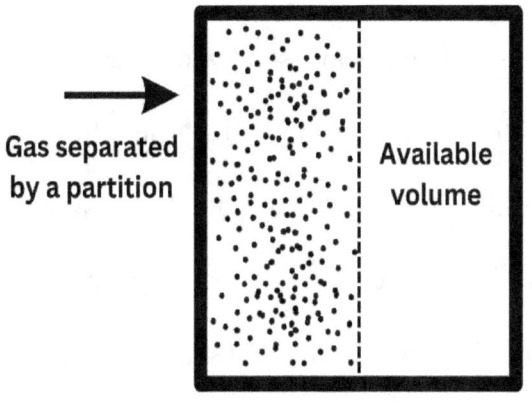

Gas separated
by a partition

Available
volume

The partition is removed and the gas files the available
volume

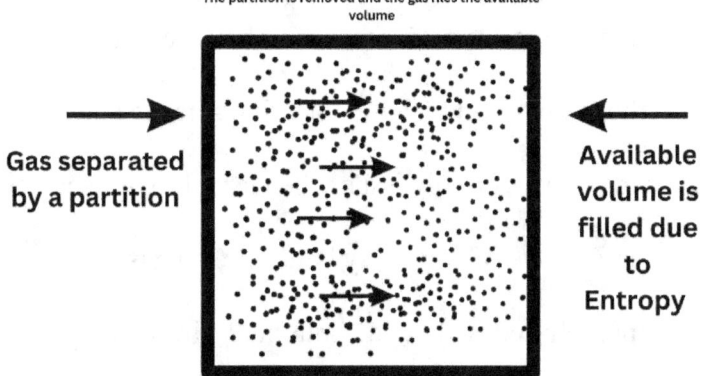

Gas separated
by a partition

Available
volume is
filled due
to
Entropy

102 Pawns In A Game

**Test it**! That's all it takes to crumble their argument of gravity holding in gas pressure. Use the scientific method and test this theory. This theory cannot be tested, this supposed hypothesis has never had to withstand the scrutiny of the scientific method which is a problem for their belief system. We in the flat earth community are only requesting that science be used when discussing the nature of our earth. There is no explanation for this other than, "gravity because we said so, and also did we mention gravity?" Those who believe in the heliocentric model are not using science and seem to have a very big issue when asked to use science in defense of what they believe. Can gas be pressurized without a container? The answer is no. Can gas law be violated? The answer is no, or else it would not be considered a natural law. Can we test the hypothesis of gravity creating an environment that we see on Earth? The answer yet again is no.

Explain to me again how I am the one who denies science.

There is no example of gas pressure without a container except within the heliocentric model. We cannot recreate the claim that we have gas pressure without a container or a vacuum without a container scientifically using the scientific model. Space, however, we are told violates this law. Entropy exists within gas law and causes gasses to operate in a certain way. Entropy is violated by the heliocentric model due to the fact we have gas that doesn't fill the available volume of "outer space." So for anyone coming into this debate stating or believing that "flat earthers" don't believe in science I am demonstrating here within just the first proof that I/we truly believe in science, so much so that I challenge the authorities on this subject to use science. To date, not one signal scientist has been able to demonstrate with an experiment anything they claim, especially pressurizing gas without a container. I'm sorry and call me crazy if you'd like but, if you propose a hypothesis and cannot test that hypothesis following the scientific method and give the excuse of using the word gravity when the answer is not known, that isn't science.

Throughout my years of research, I have had to toss out everything I have learned since I was a child and relearn new information. Sifting through that information and determining which information is good and what information is bad. What information can I verify and what information do I have to be careful with? What can I confidently hang my hat on and what information will I need to throw out with the garbage? This is a time-consuming process that takes years but some people are content with not looking into ideologies that may challenge their entire belief system. Things that make us uncomfortable are very hard to accept. New thoughts that we have never even considered are even harder to consume. My challenge to you is to become comfortable with being uncomfortable. You will adapt and overcome the years of conditioning you have endured. I can say this with absolute certainty because I was in this situation just a few short years ago. I'll admit that this was all a shock but my soul wasn't completely demolished when I realized the true nature of our earth due to my already being a "conspiracy theorist." Flat Earth is a big one because it challenges the very nature of what you thought you believed. To tell someone that you have been lied to about essentially everything for their entire life is difficult, to say the least. Yet for some remarkable reason when you challenge someone's belief in the shape of the earth, something interesting happens. Some level of fear settles in their mind. Think of the scene from Matrix when Neo first wakes up, he almost doesn't want to believe it and he admits to being afraid. Which raises an interesting question: what is so scary about being awake? I believe the answer is simple... we are taught to be afraid.

As a child you learn colors, numbers, letters, and words, and we live on a spinning ball. I am challenging one of these core lessons that has been ingrained in your mind since you were an adolescent. If you didn't major in physical science(s) after high school, this "knowledge" of the shape of the earth and the universe begins to unravel a belief that is buried deep down in your subconscious that is frightening because they begin to see what is on the other side of the red pill. The awakening is truly unsettling at first and it takes time. You begin to think, what else have I been lied to? What else is no longer part of this place that I thought was my reality? Does the moon reflect sunlight? Is the Moon a physical object? What is the Sun? Is the Sun a burning ball of gas? I thought there was no gas in outer space. How far is the Sun? What information is being hidden from us? This angers a lot of people who thought they knew everything or at least thought they hadn't been lied to their entire lives. If I can quote the late great Charlton Heston, from the movie *The Network*,

*"I want you to get mad... I'm a human being God dammit my life has value! I'm as mad as hell and I'm not going to take this anymore!"*

You should be mad, and your life does have value! Demand more, demand the truth, do not be afraid, and be willing to seek things that have been

108 Pawns In A Game

hidden on purpose by those who are playing a chess game with your life!

---

# *Science versus Scientism:*

This portion of the book is going to be challenging for some to understand. I will do my best to break down this information as simply as I can with illustrations that I hope will help. This last section of the book is the part that makes this book unique from others in the sense that I believe in science and these key points raised are usually the biggest pieces of evidence for people beginning to understand the truth of the nature of the earth being flat. All are based on science and are repeatable using the scientific method. You would not believe how many people have debated me on this topic yet have major issues with using the scientific method in testing the theories to which they cling so desperately. This is the difference between science and scientism. If you claim to use science as a means of answering questions of the unknown, yet

you then refuse to test your hypothesis, without a
doubt you are denying the very core of what we
have been taught science is and should be used for.
As Albert Einstein has said, *"If the facts don't fit the
theory, change the facts."*

---

# _The horizon is apparent, not actual:_

# _Mathematics will be their downfall_

The term horizon is defined as the _apparent_
intersection of the earth and the sky to the observer.
The key italicized work there was apparent. The
horizon we see whether we are on land viewing
across the plains of the Midwest (where I am
located) or on the beaches of California, is and
always will be flat, horizontal...wonder where they
got the root word from? The horizon is an apparent
location, not a physical location. This is a pivotal
distinction that must be declared and is a huge issue
for the heliocentric model. An apparent horizon
isn't physically measurable and if the earth is
spherical in shape or an "oblate spheroid" it MUST
I repeat!!! MUST have a mathematical
measurement to calculate curvature. There needs to
be a physical location where the earth is curving

and doesn't change, it doesn't move, it doesn't reflect. This is the case since in the heliocentric model, we are given actual numbers to measure the earth's size.

### *"Mathematics will be their downfall!"*

According to the earth curvature calculator the following is stated:

"The earth curvature calculator yields the distance between yourself and the horizon. There are only two values needed to solve this, namely the level of your eyesight or the distance between the ground and your eyes and the Earth's radius. Enter these values into the curvature equation: $a=$(square root) $[(r+h)2-r2]$. Well, there is an issue within the formulation of this equation. The calculation calls for an (r) value or a radius value. Radius is defined as *"a line segment that joins the center of a circle with any point on its circumference."* This is a logical fallacy referred to as the "begging the question fallacy." Why is this a logical fallacy? Well, the model they are using is assuming the (r) value. The shape of the earth is what is in question and they are giving a value based on their assumption the earth is an oblate spheroid. The (r) value is presupposed in the question and embedded in the answer. This is an illogical equation to use when determining the "curvature" because it uses the radius, which is assumed based on how big the earth is, or believed to be. Let's break this down a little further as we use this equation against the argument itself.

### *"Mathematics will be their downfall"*

Throughout my years of research, I have stumbled across many different experts in this field, one of which is a phenomenal YouTube channel that does a daily flat earth debate open to all who would like to engage in the conversation. Much of the evidence that I am laying out in this book (amongst others is credited to those gentlemen and their research. "The Black Swan," is an absolute mind-blowing observation that destroys the curvature calculator and the belief that there is curvature on the earth. Based on the heliocentric model, the earth if it is a sphere, with a radius of 3959 miles in circumference, the distance of the horizon measurement <u>MUST BE NO MORE THAN</u> 1.225 times the square root of the observer's height in feet. This can seem confusing however we have numerous instances of the horizon being visible further than where it mathematically should be according to their (heliocentrism proponents) own calculations. This argument is formulated as a modus tollens. If (p) then (q) if not (q) therefore not (p). When applied to the above argument it reads as follows: if the earth is a sphere with a radius of

3959 (p) then the horizon distance measurement MUST BE NO MORE THAN 1.225 times the square root of the observer's feet (q). According to the black swan argument (see illustration below) the second platform is visible in front of the horizon which is farther than the required distance according to the heliocentric model of the Earth having a radius of 3959. Therefore, not (p) therefore not (q). Based on just this one example and there are many others, the argument of the earth's radius being 3,959 miles due to its mathematical requirement used by their calculations either the earth doesn't have a radius (which I have already demonstrated is presupposed) or the horizon is apparent and not an actual location. If the earth is an "oblate spheroid" with a radius of 3,959 miles, the horizon must be a physical location, which I have just demonstrated is not the case. This is a huge problem for the heliocentric model because math is an actual applied discipline, it can not be altered by an outside force in any way. If you change the numbers the outcome also changes. The earth cannot be an "oblate spheroid" with a radius of 3,959 miles since the horizon (based on the math provided by the heliocentric model) the horizon must be a physical, measurable location that cannot move based on the equation that I provided in the above paragraph. Understandably this is the 4th time I have stated this groundbreaking discovery but you must understand this fact before we move on to the next section of scientific proofs against the heliocentric model. Many times when this argument is raised the number one rebuttal is "refraction." Refraction is the change in the direction of a wave passing from one medium to another or from a gradual change in the medium. This is stated as an explanation of why the horizon is seen further than it should be. But once again I must state that according to their model the PHYSICAL LOCATION OF THE HORIZON CANNOT CHANGE DUE TO THE MATH GIVEN BY THEIR MODEL.

For those of us in the construction industry this should be the nail in the coffin, (it was for me). Let me start by asking a rhetorical question. Can you find plumb and level anywhere on Earth? The answer is of course, yes, and this is ubiquitously true. It does not matter your geographic location, whether or not you are north of the equator or south of the equator or on the equator you can find it. Remember the above-stated fact? Math is their downfall.

Now that we know that plumb and level can be found anywhere on

Earth let's look at a simple application to prove there is absolutely zero curvature on the Earth. If we took a stretch of land say 2 miles long with no topographical obstructions and we plumbed up a steel i-beam so it is perfectly level and we did this in a line the length of the two miles so that they are all lined up. Then take a beam and level it horizontally on the vertical i-beams and use a spirit level so that we would have a level baseline. On the bottom of that steel i-beam which is plumbed vertically we ``forced the line/level" of a secondary horizontal i-beam but instead of using a spirit level, we used a rectalinator device, such as a speed square. We do this the length of the 2 miles there should be a measurable distance from the top horizontal i-beam versus the bottom i-beam if the earth truly does curve, (approximately 32 inches). This calculation of expected curvature can be checked and verified using an auto-card. Yet we must ask ourselves, at what point does a steel i-beam bend/curve with the surface of the earth?

Don't be fooled by scientism I would say the distance at two miles would be negligible because the earth is so massive remember they gave us the measurable distance to their horizon on the make-believe spinning ball. If it curves it would be measurable because we are using math with a control data point of the initial spirit-leveled horizontal i-beam. The image below should help you understand this example.

---

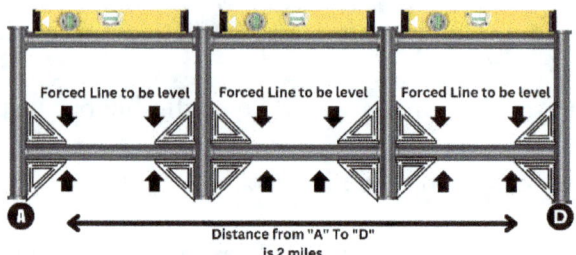

Base line Level using a spirit level

Forced Line to be level   Forced Line to be level   Forced Line to be level

Distance from "A" To "D"
is 2 miles

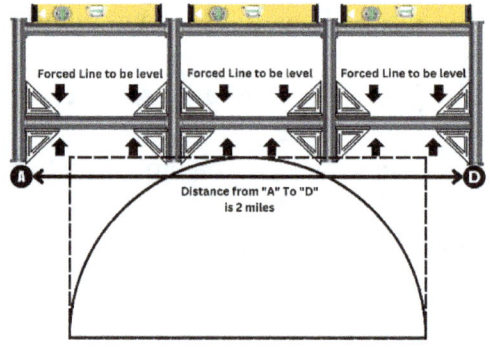

Base line Level using a spirit level

Forced Line to be level   Forced Line to be level   Forced Line to be level

Distance from "A" To "D"
is 2 miles

What does this all mean? Well in reality we know three things. Steel i-beams do not curve, plumb, and level exist ubiquitously on Earth and the earth is a motionless flat topographical plane.

This original idea was not mine and I must give credit to the man who devised this brilliant idea for a test to demonstrate that the earth doesn't curve. Brian Mullen, a structural engineer, bravely made a video demonstrating this very test I laid out above. Thank you, Mr. Mullen.

---

123 Pawns In A Game

# CHAPTER SIX

# The Lack of Coriolis

### The Coriolis Effect

Defined as "an internal or fictitious force that acts on objects in motion within a frame of reference that rotates concerning an inertial frame." Named after Gaspare-Gustane de Coriolis, the effect makes things (like places or currents of air) traveling long distances around the Earth appear to move at a curve as opposed to a straight line. Explained simply as different parts of the earth move at different speeds. The claim here is that the spin of the earth on a reference frame appears to drift objects traveling within the atmosphere to curve or alter the path of the trajectory. The Coriolis effect is claimed to affect storm rotation depending on the storm's apparent location in regards to it being in the Northern or Southern hemisphere. To note when we say the word "hemisphere," we are begging the questions within our language, this is how deep the deception goes. Storms in the north spin counterclockwise and storms in the south spin clockwise, or so we are told. Are there any experiments that prove the motion of the earth? Well, heliocentrism believes they have one.

When we ask the question about helicopters hovering if it is possible for, let's say the continent of Asia to swing by underneath the hovering helicopter. We are laughed at and ridiculed for the question. "NO! Of course not what a silly thing to ask, the earth is attached to the atmosphere duh stupid flat earther!" Yes, I have heard this before. "Oh ok, well then I guess bullets aren't affected by the supposed spin of the earth then are they?" OOPS...

You see they are wanting to have their cake and eat it too but unfortunately, the questions we asked and the answers aren't adding up.

## Foucault's Pendulum

It should be called the "FOOLS-CULT Pendulum." We can no longer use Foucault's pendulum as evidence of the motion of the earth! Those aren't my words either, those are the words of the Heliocentrists big daddy himself, Albert Einstein. The actual quote was,

*"The flattened figure of the earth or Foucault's pendulum can no longer be utilized as proof of the earth's real rotation. In relational mechanics, both facts can be equally explained with the frame of distant galaxies at rest (exerting a gravitational force on bodies at the earth's surface) while the earth rotates relative to this frame, or with*

*the earth at rest while the distant galaxies rotate around it exerting a gravitational force on bodies at the earth's surface. Both explanations are equally correct and yield the same effects. It then becomes a matter of convenience or of convention to choose the earth, the distant galaxies, or any other body or frame of reference to be considered at rest."*

In short, what Einstein was saying here was that if the bodies in the heavens are orbiting around us on Earth and the Earth is not moving the same results would be seen within our reference frame as if the Earth was spinning and the bodies in the heavens were at rest. This (Foucault's Pendulum) cannot give us a definitive answer as to whether or not the earth is moving or if the heavens are moving as the effect would be viewed the same. Of course, in chapter eleven you will see the demonstrable evidence of star trails which disproves the spinning, wobbling, tilted claim of our earth. The sentence is very telling when it comes to scientific evidence vs.

dogma.

---

129 Pawns In A Game

# CHAPTER SEVEN

# The Cable That Rules The World

Fiber optic cables are arguably one of the most important inventions of the 20th century. Carrying voice, sound, and data signals at the speed of light across the world has revolutionized communication. What a lot of people are unaware of is the importance of fiber optics and how it connects the world through its means of transportation. 99% of all worldwide communication comes from fiber optics in what is known as "underwater sea cabling." That is correct, 99%, which oddly sounds like 100%. Fiber optics works in two forms, single mode or multi-mode. The singular core of each fiber is either glass or plastic in which a signal by a beam of light is sent at the speed of light across the ocean through these cables and received on the other end. Incredibly faster than copper signal transmission and light years ahead of wireless signals. This technology has been available and widely used since the mid-1960s but recently has gained popularity. The glass or plastic tubing that is used as the core of the fiber is so small it is measured in what is called microns with a standard diameter ratio of 50 microns/125 microns (10-6 meters) and 62.5/125. That is about the width of a human hair. Single-mode fiber is used for longer transmission of data as it has a lesser attenuation loss during its travel. Multi-mode fiber is used for shorter distances as it will have a higher light loss during its transmission however it can carry a higher concentration of information. Light scatters so single mode allows a higher concentration of light. This cable links all continents together and sends information at the speed of light through a remarkable medium of glass so small that it has to be placed on a black mat to be truly seen. This is the cable that rules the world.

Why is this important you might ask? How does this prove the world is flat and motionless? Well in short it doesn't prove that the earth is flat or a globe. It has nothing to do with the description of the earth at all, however, it does lay the foundation for disproving the claim that "satellites' ' are used for any form of communication whatsoever, since satellites exist, sort of In all of the schooling I did for fiber optics never once was there any mention of "satellites," used as a form of communication. One it is illogical to think that it would be used as they do not exist in the way we are told they do and two their communication ability is worse than dial-up from the 90's. A cable that is dropped in the ocean can send signals at the speed of light, instantaneously. Sending a signal from a copper backbone communication system to a land-based tower then bouncing that signal off the tower to a supposed "satellite" traveling at thousands of miles per hour and then sending it back down to earth to another land-based tower and arriving at its final destination through a copper wire is laughable to say the least. Let's not forget all the supposed motion and potential blockages of this signal. Not only is this an impossible task that is claimed by the elites, it's much much slower and would cost a tremendous amount of money.

Satellites aren't necessarily fake but they are cartoons. Every picture you can find on the internet of a "satellite" is a cartoon or a CGI rendition. They are not used for communications due to the use of fiber optics as mentioned above. Arthur C. Clark, another science fiction writer, plays a huge role in the creation of satellites and how society views them. Mr. Clark wrote many sci-fi stories and one in particular he describes a communication device that floats around in outer space, pulling its power from the sun. His idea gave birth to what is now believed as truth. The material that satellites are supposedly built of cannot withstand the temperatures we are told exist in the thermosphere. Kevlar which has a melting point of 930 F, aluminum 1220 F, and graphite-fiber polynya does not melt but turns into gas at 6,254 the thermosphere temperature is 2,730. However, once again magic happens; the claim is that satellites can survive in this environment and don't melt because it isn't that hot. Because space is a vacuum "low pressure" system there is no atmosphere so the temperature isn't hot, but the temperature according to their model is very hot, hot enough to melt all of the material the satellites are made out of. Bippity

Boppity Boo! One scientist made the outlandish claim that when a satellite breaks in outer space they just send up another satellite to fix it. Silly. That is one of the most ridiculous things I have ever heard of. As of March 2021, there are said to be 5,779 satellites in space. I have also read reports that there are upwards of 17,000 satellites in space when you encompass every country that has a "space" program. All of these are orbiting perfectly in the sink yet never colliding and when any live videos come in from the ISS no satellites are passing through the camera frame or colliding with the ISS. No satellites ever pass in front of the moon, no satellites ever collide and NOTICEABLY knock out television or radio. See how many photos that come up are actual photos and not cartoons. What you should be seeing in 2022 are thousands and thousands of photos of actual satellites in space. Not thousands of cartoons and a handful of "real photographs" of satellites. Mr. Clark developed the idea of a satellite in 1945 that would work based on a "geostationary model," in other words, the satellite would be locked with the earth's rotation. This of course was 20 years before a "satellite" was ever developed. So let's set the record straight that TWO science fiction writers predicted the future of "space" technology and one was in part responsible for the creation of NASA. If I could insert a laughing face emoji here I would. Do people build things called satellites? Of course. Someone is making a bolt that they are told is put into a satellite which they believe goes into outer space and transmits Sunday's football game to their television screen. But this is the beauty of compartmentalization.

137 Pawns In A Game

# CHAPTER EIGHT

# We Landed On The Moon?

*"Really?? That's great!! We've landed on the*

*moon!"*

-Lyod Christmas

The moon is a great mystery whether you believe in

a spinning ball model or know the earth is flat. A

great light in the sky to rule the night that appears to

be physical. Is it even Terra Firma? Do we know?

Yet we only see one face of it. Scientists will lead

you to believe that we do see the dark side of the

moon but only during a "new moon" phase, which

incredibly, if you can believe this, is when we can't

see the moon. So let's break that down further. You

can see the side we can't see when we can't see the

side we can't see which we are seeing when we can't see it. Science baby!! The reason we are given why we only see one side of the moon is because of something called "tidal lock." Nevertheless, believe what you will about the moon it is mysterious.

Let's talk about the Tycho crater for just a minute. If you aren't familiar with the Tycho crater on the surface of the moon it is the largest crater that can be seen on the moon's face. The image above shows an astonishing photo of the moon and the visible Tycho crater. Discovered and named the Tycho crater after the Danish astronomer Tycho Brahe. Scientists have dated the crater at around 108 million years old. Not sure where they got that number from. Maybe they have just picked a big number that sounded super "sciency." The Tycho crater is claimed to be 53 miles wide with an overall depth of 15,700 feet. The moon itself is claimed to be over 230 thousand miles or 30 earth lengths away, roughly 40 times smaller than the Sun and about 25 times smaller than the Earth according to the heliocentric religion.

Let's do a little thought experiment to see where the math of the heliocentric model begins to unravel. The farthest distance humans can see with the naked eye at sea level is roughly 6-10 miles. We can see up to 30 miles on a clear day with no "atmospheric" refraction, air impurities, moistures, haze, smog, etc. in other words, near perfect conditions. So if the furthest we can see in an almost perfect environment is 30 miles, do you think it is even possible to see the Tycho crater on the moon? It is only 53 miles wide on the surface of the moon and is said to be **230 thousand miles** away?! How about a visual demonstration of this from "Google Earth."

---

The first image is taken from "Google Earth" is measures a distance from Kansas City, Missouri to just outside of Higginsville, Missouri at exactly 53 miles. The next photo shows a zoomed-out replica of that same line representing 53 miles wide at 8,000 kilometers or 4,970.9695 miles. Can

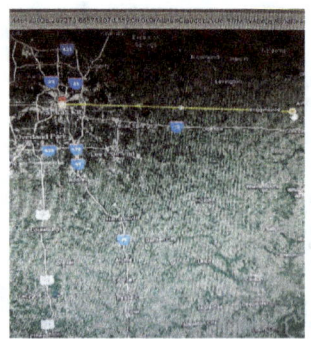

you begin to see the issue here? I still have 229,029.0305 miles to get to the supposed distance to the moon. Either we all have X-men-powered eyes but only

when we look at the moon *OR* someone is lying to us. Let's not forget also that according to the heliocentric model occasionally throughout the year the moon is even further away at roughly 238 thousand miles. I smell something fishy and I don't think it's the moon that is made out of cheese. (That's a joke, relax globers.)

If we look at history through the lens of that of a detective we can begin to unpack strange anomalies that can't be explained by official narratives. Perhaps the elites do not want us to know the truth and they truly do know what is happening in our world. For example, do we truly believe that slaves during the reign of Pharaoh built the Great Pyramid of Giza? The Great Pyramid of Giza is said to have been built between 2580 and 2560 B.C. in honor of Pharaoh Khufu. Listed as one of the 7 wonders of the world it stands at roughly 455 feet but due to erosion over time stood even higher centuries ago at almost 500 feet. An estimated 2.3 billion bricks were used in its construction but mind you, these are not your standard masonry bricks of today. Limestones were used for its construction and scientists estimate they weigh anywhere from 2 to 50 tons a piece, to this day scientists don't know how these stones were transported to the location and lifted during construction remains a mystery. Why is this important? Well, history is told by the victor, and with the United States claiming it landed on the moon in 1969 we will examine just how much of this story is fabricated. These narratives are formulated to create a worldview that fits into the heliocentric model.

In July of 1969 Apollo 11 was the first mission that supposedly landed on the moon. Well, technically not the first I suppose it would have been whoever took this photograph right? "One small step for man, one giant leap for mankind." A famous phrase by the legendary astronaut who took man's first step on the moon. Still, wondering who took the photo though? The classic and comical response to this was there was another lander that was sent out from the original lander that had a camera attached to it that the astronauts could control remotely with a controller. Oh boy... Do you know what the astronauts had at their disposal on the moon? Grown men in diapers walking around, playing a bit of golf, and driving around a buggy! A lunar buggy that somehow they attached to the side of the lander? What kind of attachment would be capable of adhering to the speeds that are claimed to be needed to "escape Earth's gravity?"

145 Pawns In A Game

PLEASE EXPLAIN TO ME HOW THE BOTTOM PHOTO OF THE LUNAR BUGGY FITS INSIDE THE TOP PHOTO OF THE LUNAR LANDER WITH ALL THE ASTRONAUTS AS WELL??!!

---

147 Pawns In A Game

# CHAPTER NINE

# The Challenger Disaster

**(This entire chapter is pure speculation)**

*"We will never forget them...as they prepared for their journey and waved goodbye and slipped the bonds of Earth and touched the face of God."* - Ronald Reagan

What year do you recall the Challenger disaster taking place? Some remember this disaster taking place in the late 80s. Others, such as myself, remember watching this live on TV during the 90s. Some remember being in school and watching it on the old TV that was ratchet strapped down to the old four-wheel cart. As interesting as this story is, it is even more interesting that this story also plays into another conspiracy that has been circulating on the internet for the past few years. The Mandela effect is something that millions of people around the world are familiar with yet this chapter won't cover that. We are going to only focus on the official story of the Challenger disaster and how it pertains to a cover-up as the shuttle was never supposed to explode.

I believe that no human is ever present in the shuttles when they launch. If you have never heard of the Challenger disaster then buckle up because this story will knock your socks off. The level of absurdity shrouding this story is uncanny.

Flight STS-5L-1 (so **S**tupid some people will believe **T**his **S**tory) took off on January 28th, 1986 reaching an altitude of approximately 46,000 before exploding in mid-air claiming the lives of all 7 astronauts on board. Who were these seven astronauts? Let's meet the cast and crew of the Challenger disaster. Captain Francis Dick Scobee, pilot Mike Smith, mission specialists Judy Reznick, Ronald McNair, Ellison Onizuka, aircraft engineer Gregory Jarvis and last but not least the first teacher to attempt to enter space to demonstrate the importance to kids seeking highly technical jobs, Christa McAuliffe. Pilot Mike Smith during the press conference before the shuttle's fateful take off is quoted as saying "...we are all looking forward to getting up to orbit and getting the secret handshake." Secret handshake huh? Not a masonic secret handshake right? An odd thing to say during a press conference right? Christa McAuliffe, the school teacher, had never flown a mission before as she was, you know a teacher, Gregory Jarvis had not flown a mission either and this was Mike Smith's first piloting flight. Interesting to note how promoted this flight was considering you have half of the flight team made up of rookies who had never "left Earth" before. A television set was rolled into every classroom during this launch to demonstrate the grave importance to children of space exploration. Unbeknownst to those watching little did they know they were about to witness something that was never supposed to happen. If you haven't caught on by now there is no "outer space." We cannot leave the earth, we are stuck here under what I believe is a dome as described in the Holy Bible as well as almost every other ancient text. Space as we are led to believe is not a vast nothingness with other galaxies, planets, and potential life, we are it. What does this have to do with a shuttle blowing up in 1986 or 96 depending on which reality you subscribe to? Well as I mentioned earlier the shuttle was never supposed to explode obviously. It was supposed to make its illustrious arch into the sky crossing over the ocean until it faded from the view of the cameras and those watching from Cape Canaveral and hours later a full crew of people floating in space was to be shown to millions watching. When the rocket exploded mid-air, live on TV it changed the narrative instantaneously. Damage control was immediately rolled out as world leaders took to podiums to give commemorative speeches about the brave lives lost on the challenger. The space program was halted for several years after the disaster and an investigation was launched into why the shuttle exploded. You see the issue here is no one goes into space, so the shuttle is launched over the ocean and then the magic happens. When the shuttle exploded live on TV and everyone saw it, now we have to commemorate 7 people who died, who aren't dead and a cover story has to be presented. This is where the story gets strange. You may be thinking to yourself, OK now I'm drawing a line here, claiming that astronauts didn't die in the Challenger disaster even though we have all heard about this or seen it on TV is absurd! You have lost your mind! Well let's examine this for ourselves, shall we?

Below we will see some pictures of the astronauts on the shuttle before its disastrous outcome and then we will see pictures of them now except one. Gregory Jarvis is the only one of the seven crew that never reappeared after the supposed death of the crew. Perhaps and this is just my wild imagination, he felt the guilt of lying to the world about what took place and was willing to talk, see his family again, tell the secrets of those in the know, so he was dealt with, allegedly of course. The captain of the crew, Francis Dick Scobee just goes by Dick Scobee now and is the CEO of a company called "Cows In The Trees."

Christa McAuliffe the famous schoolteacher who lost her life that frightful day goes by her first name Sharon McAuliffe and is a law professor at Syracuse University. During her NASA days, she was using her middle name, Christa. Judy Reznick, mission specialist, didn't even bother to change her name and teaches law at Yale. The pilot Michael Smith also didn't have the guile to change his name and is a law professor at the University of Wisconsin. Mission specialist Ronald McNair has an IDENTICAL TWIN brother by the name of Carl McNair.

Ellison Onizuka...ALSO HAS...HOLY SANTA CLAUS SHIT an identical twin brother by the name of Claude Onizuka. Wouldn't you know two twins aboard the one shuttle that just so happens to explode live on TELL-LIE-VISION, which sets up a relatively easy cover story for the surviving members?

154 Pawns In A Game

This of course is completely a theory and holds no weight whatsoever....remember this is just a theory.

155 Pawns In A Game

156 Pawns In A Game

157 Pawns In A Game

# CHAPTER TEN

# Star Trails

Most people who believe in a globe model don't realize that when they discuss the sky they mention the stars rotating above us; which of course I would agree with. The fact that stars leave symmetrical patterns above us in the heavens is proof that the Earth is fixed and stationary. If the earth were a spinning ball or an oblate spheroid that rotates at 1,000 mph at the equator, wobbles, and orbits the sun all the while the entire galaxy is shooting through an infinite and ever-expanding universe, images below would not only be impossible but the symmetry of designs of would paint the sky evening after evening which would have left our ancestors lost when attempting to navigate using the heavens above.

Perfect symmetrical circles around the northernmost (fixed star) Polaris. Although the North Star does move, the movement is so slight it is almost impossible to see with the naked eye and even more difficult to catch without the proper training. Now imagine if you will that without this model of the flat earth we can observe things in the sky such as stair trails, symmetrical patterns, and constellations that reset to the exact positioning each year, with no motion felt with our vestibular systems, are all things we can see and feel. Yet the globe model requests that you do not trust your senses. "We know you don't feel motion but trust us (insert lab coats here) we are moving." Then imagine that you are told that the theory of the earth being a globe has wonderful explanations for all the things we do see, star trails, gas pressure without a container, seeing too far past the horizon, etc. but unfortunately, these phenomena cannot be demonstrated or recreated in any way with any model or

experiment. That of course is okay though because to demonstrate, mass attracting mass is an example. they would have to create a model so big it would be physically impossible. Does this sound like science? Does this sound like the application of the scientific method?

Now let's

discuss further the complex nature of star trails. The issue has arisen with the "southern celestial pole star" being visible directly opposing the Northernmost star Polaris with apparent opposite facing star trails. Let me start by giving you a quote by Dr. Samuel Rowbothum, *"If the Earth is a sphere and the pole star hangs over the northern axis, it would be impossible to see it for a single degree beyond the equator, or 90 degrees from the pole. The line-of-sight would become a tangent to the sphere, and consequently several thousand miles out of and divergent from the direction of the pole star. Many cases, however, are on record of the north polar star being visible far beyond the equator, as far even as the tropic of Capricorn."* -Dr. Samuel Rowbotham, "Earth Not a Globe, 2nd Edition" (41)

With this above image, you should understand how impossible it would be (if the earth were a spinning ball) to see the North Star if my position was anywhere below the equator. Yet countless times the North Star has been identified as far south as the Tropic of Capricorn.

## Celestial Navigation

Celestial navigation is a topic that I am certainly not an expert in but will end this chapter with this. A sexton, like the one pictured left, operates on the premise of finding an acute angle between a celestial object in the sky with its relative position above the horizon down the arch second. Essentially you would like your horizon mirror up to the horizon (that should be self-explanatory) which would establish a horizontal baseline. You would then focus your second mirror onto the celestial object by changing the angle to the object until it was focused at the sextant which would then give you an angle of degree. With a bit of math you can find the GP or ground position of the celestial object you are looking at, this of course could be thousands of miles beyond the horizon. This is how celestial navigation was and has been used for hundreds of years. This is a very accurate instrument, that allows for minimal inaccuracies because it is used to find an angle, and with basic geometry, we know that you cannot find an angle with curved lines, let alone a reference frame that is spinning, wobbling, orbiting and shooting through an ever-expanding space.

---

164 Pawns In A Game

# CHAPTER ELEVEN

# The Conclusion

We have finally reached the end of this journey, but hopefully not for you. If you are reading this book and made it this far I challenge you to do your best to continue your search for truth. The "Flat Earth theory," should no longer be considered a theory but an unproven fact in your mind moving forward. You will come across information that will make you wonder, perhaps a question will be asked about how something works on the flat earth that will stump you, however, do not be discouraged the answer is out there. What tends to happen with this topic is that people will expect you to have all the answers to how the universe in its entirety operates perfectly and if you cannot supply that answer they will assume your original premise to be incorrect. This is a tactic used often to discredit the search for the true nature of Earth, especially if this is your first experience with this topic. DO NOT FAIL FOR THIS TRICK! Keep digging and find the answer because it is out there. What you have to remember is that people who will question what you say are hoping you are more ignorant than they are about the topic at hand, especially with flat earth.

This is a topic that took me 6 years of research before I finally said to myself, "Yes the Earth is a flat motionless plane." But the next question was the hardest one to ask me and it is one that I search for the answer every day; "How do I prove this?" This is a question which you should leave this book with. I do believe that I have left

166 Pawns In A Game

enough evidence for you to not believe that the earth is a flat motionless plane but to KNOW. How do you disprove curvature? Test for curvature. Would it shock you to find zero curvature in your measurements? Long-distance observations, gas pressure without containment, finding plumb and level, the list goes on but the ball is no longer a fact we now know is a lie.

I want to thank you so much for reading this book and supporting my endeavor of publishing my first written work on this matter. I can assure you there is much more to talk about on this topic. I did not even mention gravity, space launches, high-power zoom lenses, spectroscopy, distant observation, time zones, flight paths, and the list goes on. There will be a revised edition with much more information added. I felt obligated to get this copy out to the truth seekers so that the world could hear this message. Believe me, there will be even more soon because the work to uncover the hidden

truths remains at large.

---

**Remember we are all pawns in a game.**

www.ingramcontent.com/pod-product-compliance
Lightning Source LLC
Chambersburg PA
CBHW070012300526
45794CB00001B/296